Parting and Enduring

Mathieu and Michael MILOUD

« It is not the strongest of the species that survives, nor the most intelligent that survives. It is the one that is the most adaptable to change »

Leon C. Megginson, about the work of Charles Darwin

Prometheus

Yes, the day will come, no matter how stubborn he is, for Zeus to fall, since he plans to mate in such a way that he will be thrust into darkness from his tyrant's throne, and so he will bring to pass the curse of his father Cronus, uttered at the time when he fell from his throne. How to save himself, only I, from all the gods, have the means to reveal clearly. I know exactly what to do and how. He can sit boldly all he wants, trusting in the reverberating might of his lightning bolt. He can shake it all he likes, but it won't help him. These fire-breathing bolts cannot protect him against this dishonorable and unbearable fall. Such an opponent is he creating against himself, and a marvel impossible to defeat. He will discover a flame stronger than lightning, a roar that will deafen thunder, and shatter the trident, Poseidon's spear, scourge of the sea, and land-shaker. In his catastrophic fall, Zeus will learn how great a difference there is between ruling and being a slave.

Aeschylus, Prometheus Bound, Translation by Marianne McDonald

Life's theory of evolution, or at least its currently most widely accepted form, first appeared in England. The fundamental idea behind this theory, is that those who live the *longest* will write the next part of the story, and since future life comes from a *maximum*, it will be more solid than present life. This conception is very similar to the English way of seeing economics: everyone as a part of society possesses a power to bend the will of others which may be measured by a *number*, an object that can be *added* and *ordered* like on a straight line. Wealth attracts wealth and poverty rejects it. Poverty doesn't count and wealth goes on.

Yet our theory of evolution must go beyond English thought, for living beings are not increasingly stable. Any comparison of stability between species living in different times can have no experimental basis and would thus be void of meaning. Indeed, on a small time period, an evolution theory with correct foundations states that, in a fixed setting, life's stability keeps growing. However, the world does not stay eternally the same and life is its own surroundings. Each weapon life invents against another life form might eventually be used to fire back when it becomes a common property. But when mankind realizes this through its own weapons, it will be too late to choose another path.

One might believe that if mankind comes to an end by its own weapons, it will have been the victim of an unlucky and unlikely fate. But mathematical theory of probability provides a clear judgment on this: every event that might occur at any time, even if the odds are against its realization at each moment, provided that its probability (which represents its possibility) does not decrease, will eventually happen. Furthermore, under these assumptions, the probability that such an event shall not take place during a time period decreases exponentially.

Mankind now has the power (though it might be deemed weak) to destroy itself. To escape this end, such a possibility must continually diminish.

Two trails can lead to this: the decline of the united civilization, which would leave it unarmed, or its schism.

Let us consider the first road, undoubtedly the easiest to follow since it is necessary: every civilization must die. Nevertheless, saying this comes down to ignoring the questions of when and how, while simultaneously hoping a potential future civilization will answer them in our stead.

Every human group started as a tribe, but since only the

greatest assemblies could survive endless war without getting annihilated or absorbed, societies have always kept expanding. These changes radically affect the individual. Those with whom he shares the land that feeds him, he cannot literally call brethren anymore. Natural instincts progressively lose their ability to propose an adapted relation to others, the lack of which is replaced by culture. *Harmony* between nature and culture wears out while *beauty*, this lie that binds both as one holds civilization in a perilous balance.

Now that mankind is united as it never was before - at least out of Asia -, every loss *beauty* sustains is no longer an opportunity for another civilization, but simply is the sign of an aging and that death is getting closer.

The only option left to prevent a disaster such as the single destruction of civilization is for it to get divided by reaching new spaces. Roman Europe already managed such a jump towards America in the XVth century. Does it still have enough strength to go even further, to another planet? This might be doubted, but if it is to come true, mankind must focus on this goal, for no thought can be neglected in order to adapt to an unknown environment.

If, after having tried, man was to fail before such a hard task, he could nonetheless send simpler life forms far from earth. No other living being could do so by itself by *small* steps, which have generated all life on earth. A *huge* push is necessary to cross such a space - and time - gap. I don't conceive how this could arise without *will*.

Man won't have an open door to other territories forever. Not only does this project require deep insight such as Man has still to prove, but mankind will also need resources that are nearly depleted.

We know very few men who possess the *will* to secure their species' future. And though everyone partakes in it, only few would claim it as their purpose. Hence, we shall try to encourage this will and goal.

We speak to you, powerful men. Aren't you tired of seeing you've won against weakness? And of *counting* to what extent? Aren't you tempted to achieve something that cannot be foretold? Don't you wish to face a challenge you are worth of and competition even greater than you? Watch the knight who rides his proud horse. Why would he have purchased it with gold if it weren't to go far? Further?

First, we will explain what we suppose are the reasons that led us to writing this book. This isn't important so we shall be brief. However, we feel we ought not to hide this to the reader, if we want him to trust what we suggest. Indeed, the influence of a will on another starts when an individual has an aim in mind. He plans how he can fulfill his will, with the assistance of others. If he can convince them to share his dream, cooperation is natural. If he cannot, he must, as the fox in the tale, deceive the crow by giving him a target delusion, so that his arrow might hit what is behind, transforming his means into the others'. But no such intention can be found in these lines. For we propose a goal, but no plan from which precise consequences may be expected. Moreover, individuals who are entitled to make decisions for many men, for whom we write this, cannot be misled. If there is doubt left concerning our motivation, after having read this book, ask yourself: who can predict what the outcome of such an effort carried out by mankind would be? It could exhaust a civilization but could also become the source from which springs a new feeling of unity among men.

The point is: where does this book come from? As authors, we are not members of a lineage rooted in a long-established tradition. We had to seek what direction we could take as various instincts contradicted themselves. Surrounded by different schools, we discovered a method that allowed honesty towards oneself and consistency. This "scientific method" starts with the interpretation of observations by hypothesis to describe how such phenomena generally unfold. But the valuable quality of science is the second step of its mechanism: a scientist will try to contradict his own hypothesis. If that attempt fails, his hypothesis is likely to be true, it is a close approximation of reality. In other words: what first seems to work doesn't always turn out so, but what has long resisted is stable. Applying such a principle to one's own ideas is nothing else than self-respect. As for the consistency we hope we have acquired, it is made of logic which we learned from the study of mathematics. The next accurate lens was the theory of life's evolution through the selection of forms according to their resilience in certain conditions. We had tools to understand some structures of the world, yet we had no pre-ordained goal. We may have set goals for ourselves, on individual levels, but they are irrelevant here. As regards our actions on a non- local scale, they could not be arbitrary since arbitrary effort is unstable and thus cannot achieve much. Our individual town, region or even country are none of our concern. *For us,* they consist in arbitrary drawings on a map. Where then could we place *our* actions? We are not universalists professing intermediate scales do not exist or are illegitimate, but nothing links *us* to them. Apart from the local scale, the only level we are interested in is at the other end: the global one, mankind or even Life.

Maybe the reader is getting bored. We shall now present the conclusions we drew from our investigation.

There are several perspectives on the question of how life can be spread in practice. The first one is that, although man's body isn't adapted to living outside earth, men's *will* enables them to build a shelter outside their usual environment.

If man was to accept that he cannot adapt out of earth, other life-forms from earth such as bacteria are more flexible, and he can select and transport them. Of course, that would make the re-appearance of *will*, necessary to continue man's work, quite long and hypothetical. Life's extinction would be but postponed.

Finally, if these are dead-ends due to a too strong connection of terrestrial life to earth, man's last resort would be to attempt the creation of machines obeying life's evolution's principles specifically adapted to another planet.

Before developing these three points, we shall discuss evolution.

Chapter 1 - Metamorphosis

Senses are finite inasmuch as they will never perceive more than a tiny sample of the world. No *essential* rule linking objects can be derived from viewing the world, only likelihood can be evaluated by relying on experimental evidence that has never contradicted theory. That is the core of experimental science. After having guessed what underlying law could describe a certain class of phenomena, one may wonder if this is merely incidental or if it is more general and hence test it in manifold situations. It is often obviously impossible to test this law in every configuration, and thereby to *prove* it. With limited information, the only criterion of truth is that the law has always been right. And honesty requires one to collect the largest possible amount of data. A theory can only be taken as valid after having been attacked. Notice that a prior condition to test a theory is that its predictions be unequivocal, so that mistakes can be detected. This falsifiability defines the experimental or scientific method. For instance, if one were to conjecture that "All elephants are green" one would quickly have to admit the error by visiting a zoo. On the other hand, after numerous verifications, the following rule is quite reliable: "If I drop an object without throwing it, it heads directly towards earth."

This method can't answer every question about the world, but it is the only one a man possesses to understand what is outside him through his incomplete perception. It only describes the world, since every law carved in its tablets is the result of gathering numerous phenomena. It can tell *how* but not why, since the question *why* looks for an intention.

Yet, this method isn't fruitless, and gave rise to the science of matter, its links and movements named physics, whose prophecies are quantitative. But we will not focus on this part of science here, rather on another; although less precise it is still significant, as the forms it studies are guided by a logic principle. Before grasping this fundamental idea, human thought had to wander, search and confirm its discovery. This story is part of history.

It all began in the nineteenth century. At that time, European society was changing at an ever-increasing pace in different fields: economy, technology, politics ... and what used to be unmovable ceased to be so. Philosophy was impacted by such transformations, and at least three major thinkers personified this new feeling: the naturalist Charles Darwin, Hegel, who built his historic interpretation on his dialectic logic system, and the singular Nietzsche. From 1831 to 1836, Charles Darwin was navigating near South America aboard the *Beagle*. He wrote down his observations about geology and biology on the islands and coasts where the ship anchors. These include notes about fossilized animals, and species still alive, such as thirteen species of tanager birds from the Galapagos Islands, which resemble each other but can still be individually identified by features such as their beaks whose forms correspond to the food they eat where they nest. After his return to England, Darwin took time to think about the innumerable living forms he encountered during his trip. Darwin wasn't the first to relate the differences between close species to their usefulness in their life condition. His contribution was to show what process could bring this about. Two elements are keys to this hidden treasure: evolution and selection. As parents are imperfectly copied in their offspring, random new qualities may appear at each generation, which accounts for originality in life and its evolution. However, novelty is not synonymous with improvement. Furthermore, living beings in which flaws accumulate will degenerate and become dysfunctional. Darwin, standing before this obstacle had to introduce the notion of selection. Each living being that perpetuates its species will have suffered hardship, escaped from starvation, predators and bred. Surrounded by many competitors, the weak stand no chance of winning. Living beings that are born with a new feature that hinders normal abilities will soon leave no trace in their species' future. Three-legged born antelopes are easy prey for lions and are eliminated. Conversely, a living form that developed a particular non-inherited strength might have a large descent to which it can be passed on. Rare though the latter type may be, it is the only one that matters. Through this, species gain stability. If through migration, two components of a species are geographically severed and stop mixing, each one will evolve independently from the other. As they can no longer crossbreed, two distinct species become defined. That is how Darwin explained the origin of species [1].

Darwin's intuition was triggered by observation, with later experiments such as digging fossils confirming life does evolve, and a huge portion of life is wiped out on large time periods. The human brain isn't used to such scales; hence it is prone to doubt such complexity can be produced by sheer randomness before selection. Nevertheless, anatomy comparisons clearly reveal genealogy.

This theory is not a mere artwork of logic, it is a masterpiece that can thoroughly sum up Life's slow complexity growth, as living beings start carrying shields they picked up unaware of their future use and take the upper hand.

Evolution theory successes didn't end with Darwin. In the middle of the twentieth century, the structure of a molecule called DNA, which all living cells contain, was described. It is made of two linear sequences of chemical "letters" (A, T, G, C) facing each other, with an A always connected to a T on the other linear chain, and a G to a C. Some extracts of this DNA - called genes - are translated into proteins, molecules that govern all the chemistry in a living body. During sexual reproduction, each parent has two versions of each gene, and randomly gives one to their heir, thus ensuring either none or the entire gene modified by an error is copied, and evolution doesn't dissolve.

We would like to concentrate on two prolific corollaries of the theory of evolution. If one puts together the fact that forms which survive are adapted to their surroundings, and that what surrounds them changes, one deduces that life must be stable. Living forms must not only be adapted to their current living conditions, but also many other close ones to which they might contact, and that many of their forebears have certainly also encountered. Plants, even though they cannot move, should not die from the first blast of wind or rain. *Stability* is the fact of remaining, to some extent, unshaken after an exterior blow. Stability is an ingredient of Life.

Secondly, we wish to emphasize that material organisms aren't the only ones to submit to such evolution. An example of a category of evolving forms, in the world of behaviours, words and thought, is that of memes, a word forged by Richard Dawkins referring to an elementary communicable human action [2]. A class of memes that can be easily relayed is that of language memes, which are *symbols*, meaning that their use immediately associates them to various other notions. If you hear someone mentioning the word tree, you might imagine a form with branches, leaves or maybe thorns, green or yellow, as in the fall. A symbol may circulate without conveying its initial meaning, that is its interaction with other memes - which is encoded in the brain. Social group borders are marked by the discrepancies in the symbols they share and by what they put inside their symbols: for instance, French and Brazilian recollections of the 1998 World Cup will not be the same. In mathematics, formalism is often efficient to avoid misunderstandings, although even mathematical formalism cannot fully transmit a geometrical idea. The existence of a symbol does not depend only on the consistency of what it represents and might survive merely because it can be easily repeated, such as a motto or a melody. Such are the memes: "voter est un droit, c'est aussi un devoir civique" or "every penny counts" which can be contested as to their truth. These memes act just like living organisms as they reproduce and can mutate as they leap from one bearer to another, assuming a new form in the terms they are composed of, although memes are often willingly created, which most living organisms are

16

not - except for GMOs. The adaptation of memes comes from their capacity to be memorized and to be taught to others. Oral memes might emerge from rhythm, rhyme and musicality. Music is in fact quite mysterious, since it utterly controls man, piercing his defences. As such it can be regarded as a burden in evolution. Since only noise exists in man's natural environment but not music, music may be an anomaly, as a remnant of the ancient perception of sound, which has not yet been replaced by a more efficient process. Unless the role it plays through songs in seduction and therefore reproduction and more generally in bringing men together outweighs the drawback on the individual.

These two ideas imply that the theory of evolution might be extended beyond its usual boundaries: if memes behave as living forms, other forms also might. *Stability* is the gate to this comprehension. Contemplating the forms we identify around us, notice they all share stability, since if they were totally chaotic it wouldn't even be possible to define them, by stating a constant relation to other forms. An elephant is a large size grey animal, and stable in that sense. Do absolutely unstable forms exist? That we cannot ask, as we could certainly not exhibit *one*, since *one* indicates unicity and therefore stability. There is nothing to marvel at if the human brain only recognizes stable forms, for they may help to repel ever-harassing dangers while a strategy may be designed to fight them.

Stability being granted to everything man calls forms, how can he discern living forms and thus generalize the principles of evolution? Evolution takes place on a stage where variation scope is limited, on which living forms must be stable. The stability of this stage itself is experimentally postulated. The stage consists in the laws of physics, which men can deduce from the empirical method already mentioned. By laws of physics we do not mean a closed domain, including areas such as mechanics, thermodynamics and so forth; rather a set of phenomena that implies many more without consequences of others unto themselves. From that point of view, life's evolution is a corollary of the laws of physics.

Up to now, living beings were treated in a most evasive fashion as to their definition. They were branded *stable*, but that is not sufficient to distinguish living forms from inert ones. A living form is one among numerous others of its species. We will say it is a *remembering stable form*, not that it has a memory in the ordinary sense that it would emerge from the human brain. Instead, here we mean that if several similar forms exist and some of them - too fragile - are destroyed, then by contrast those that subsist are armoured against harm, and the armour we call memory. However, we haven't yet dug deep enough to examine the roots of Life, for some remembering forms could hardly be considered alive: rain drops, planets of the solar system and such. A last property of living forms is that they can replicate, they are *copiable remembering stable forms*. That is how we define life, without mentioning any specific process linked to earthly biology. We do not care if there is water on a planet to determine whether life can flourish on it, for water is not part of our definition of life. Now what can be concluded about general living forms? How can evolution theory be stated in this context? Stable forms never being absolutely so, they tend to slightly drift from their original course, and evolve. Remembering stable forms are then selected. Nevertheless, seeing that any individual form will eventually deteriorate and vanish, for the newly selected feature to outlive the form it is part of, the latter must replicate, and must therefore be a copiable remembering stable form, in other words it must be alive. These three phenomena, evolution, selection and copy permit life's complexity. That is how we rephrase evolution theory for every copiable remembering stable form.

Notice that we haven't pointed out any stable form that isn't a remembering stable form. And they seem quite exceptional indeed, but here is one worth naming: Life itself that is unique therefore not *remembering* despite its *stability*.

Evolution theory has now been suitably presented, but we wish to further insist on an aspect of copy that is still implicit in this text. This aspect is *time*, in which evolution takes place. Experimentation doesn't give many clues about the nature of time - and we cannot even suggest what could be expected from such speculations. The question we ask is merely that of the orientation of time, assuming it is a sort of continuum shaped like a line, in which events occur relatively regularly. We can give four ways to fix the orientation of time: three are biological, and one is physical. We wish to show their mutual position isn't purely accidental, to understand them better. The first observation to be made is that Newton's second law, that governs material movement is invariant by reversion of time. There is no privileged direction of time in mechanics *a priori*. Nevertheless, no one will deny there is one and we are certain there is a past and a future. Spontaneously we would probably say the past is what we know, and the future is what we cannot say for sure. It is therefore our memory - here in the classical sense - that provides a first biological definition of time orientation. There are two other such ways to detect a time arrow in life: first, as time passes, life's complexity - its number of defences - increases. Second, replication creates more living forms in the future than in the past. Before dealing with the physicists' findings on the time flow, we will try to merge the three first definitions. As each of them is valid for a certain system - physical world, organism, learning organism - we will attempt to show they agree at a common level. By applying the definition based on copy, let us deduce the two others. Let's first acknowledge that all organisms on earth are surprisingly analogous in respect of their DNA, which is read in a quasi-universal language and prescribes what proteins to build, therefore being responsible for all bodily chemistry. This cannot be the result of random convergence. Currently living forms must still convey the legacy of an outstandingly thriving unique ancestor. This ancestor, in order to be alive, had to replicate, which means at some moment there was one form, and at another instant, there were multiple. Having copied its attributes, it copied its copying direction, thus synchronizing *replication orientation*. At each generation, new characteristics can appear. If they favour stability or reproduction, they will occupy a stronger proportion after their replication that will have been more efficient than that of their fellow competitors.

And *complexity rises* in the same *direction as copy*. As for the memorization, sexual reproduction is a bottleneck through which even multicellular organisms must pass as once unicellular. At that point such organisms are incomplete and will tend to completion. Even at their birth, these organisms might not be entirely autonomous as their parents can still assist them for a while. However, since they must be able to acquire independence as they should surpass their parents' existence, memory is logically selected to absorb information from birth to death. We trust this explanation is convincing. We will now focus on the connection between the biological and physical perspectives of time.

Physicists realized that if one lets a system freely evolve, it spontaneously slides towards homogeneity. If you mix a cup of water and a glass of wine, the colour of the liquid will gradually turn smooth. This is a manifestation of time producing a quantity called entropy which supposedly stands for disorder. Let's try to justify this. A prerequisite is to introduce a fundamental tool. *Forces*, ever since Newton, are used to describe the influence of one body on another. A force deviates or alters the velocity of matter. However, that cannot be measured. Let us explain: if all the forces acting in the universe were to impose the same movement to every object, the effect would only be virtual, and not tangible from inside that movement. What is palpable is the different effect forces have on different components of the world. For instance, a force that impels only magnetized matter has visible consequences. Similarly, a force that affects different zones in space and time differently makes matter converge or diverge.

21

In every case, *forces separate*: they separate different kinds of matter or else, like gravity, matter from vacuum by aggregating it on a convergence point. On the contrary, if one forgets these forces, which as a first approach cannot be considered absurd for gases, the stable form is *homogeneous*. That does not mean to imply that other forms do not emerge, but their instability rapidly puts an end to them and when the stable form takes over, it prevails for long. In particular and to repeat, when you pour wine into water it will quickly assume the aforementioned stable homogeneous colour.

We use the word *form* that denotes human judgment on purpose. Indeed we do not back absolute Galileanism, which would have all laws of physics invariant under a common category of transformations (such as that named after Galileo or the relativistic improvement that was later discovered by Lorentz and Poincaré). Thermodynamics is a practical topic, initially studied at a human scale in order to waste less fuel in engines, therefore it only addresses what men can control and measure, excluding ipso facto Maxwell's demon that does not play by such rules. Hence, thermodynamics should satisfy invariance after exchanging *human* witnesses rather than invariance under permutation of Lorentzian or Galilean reference frames. We do not believe it is relevant to speak of a universal quantity called *entropy* that can be totally ordered in a mathematical sense. Of course, one can take advantage of an expression of entropy in special simple situations such as a system composed of several ideal gases to foresee what the outcome of some experiment may be, but we have already discussed that living beings also have an entropy. To that we will invariably reply: So what entropy do you assign to this fish? If you cannot say, then you are in no way an empirical scientist." A human observer can *recognize* many forms, but that should not mask that shapelessness is far more ordinary than forms: in a picture, any arbitrary portion is not a form although it may contain forms. In fact, a form must be consistent, involving a force that separates it from the rest. How come? All voluntary living beings, and among them human beings have their will selected to react to stimuli from the outside world that affect survival regularly enough to require an answer. This leads to the condition that forms must be non- homogeneous - for what is constant deserves no attention - and have a unity and a regularity that is maintained by *force* and allows to count them. That is what we term a *form*. The second principle of thermodynamics precisely claims that if one cancels forces in a system, as when one removes a barrier between two fluids, it tends to homogeneity. But when forces disappear, we explained that shapelessness predominates, and as forms are relatively rare, the chances to fortuitously glimpse a mirage of form inside it are low.

So where does the coincidence between this orientation of time and the biological come from? This is concealed in the possibility of performing such a deed as mixing wine and water. In the duality opposing *separation* to *homogeneity*, life supports separation, for copying and growth are executed by separating inside the cells chemicals collected outside the organism. Before you could blend the two liquids, the wine must have been extracted and separated from the other elements of the soil by the grapevine. One may argue that our example is too directly attached to life. However, we chose this only to stay clear.

Indeed, if in many circumstances the understanding in terms of form may decrease, with no intervention of Life whatsoever, apart from human observation, this does not take place independently of Life. When raindrops fall in the ocean, a cloud fades out, but it exists at some point and although life didn't produce the cloud, they both were separated from earth by the sun beams and returned to dust by gravity which is the common destiny of all forms on earth that receive heat and then lose it. And chaos succeeds order when what sustains it is gone as death follows life. The second principle is pessimistic in its terms as it concentrates on the end rather than on the story.

The role of evolution in our perception of time has now been exposed. This completes and clarifies our understanding of evolution.

We now wish to explore a tiny portion of the ocean of consequences that flow from the theory of evolution. Evolution must always be kept in mind as the origin of everything man thinks, feels and does. All that we have been writing in this book is based on a logic, and this logic is worth using for it enabled previous living beings - humans and others before them - to survive and reproduce. One may regret not being able to master ideal logic, but there is no choice and these are still solid grounds. We endeavour to filter memes through logic to eliminate those that last only through their capacity to be memorized and recited despite their faltering truth. How can reasoning make genes accord with memes? The human brain starts by *intuiting* what is correct by assembling different forms to obtain a new idea. If you stare at an object alone, and then a second one comes into sight, and now you behold a third, and so on, you might notice these amounts form a series and thus apprehend integers. We do not assert that integers are not *instinctively* employed by men - that is that individuals don't learn this notion as evolution already thought them - our purpose was merely to illustrate the first phase of knowledge. This intuition is nearly what Plato used to call right opinion [3], although he believed it arose as reminiscence from a past perfection of the soul. *Inspiration* from a hypothetical fountain that dispenses ideas has also been suggested by poets. But as they cannot designate its location, this only drives us away from a solution and we will stick with intuition.

Nevertheless, intuition can make mistakes and its results must be sorted according to their truth by reason and experiment. In some cases, intuition may even be bypassed as pure reasoning can lead to an unexpected deduction, as it happens in mathematics, before a theorem has been proved or disproved.

Evolution theory also affects our interrogations about the world, substituting How? for Why? by abandoning the *ultra-complex God* the nature of whom we ignore and replacing God with an elegantly simple account of complexity we understand enough to wager what type of living beings might exist: they should not be perfect but only operational with the available means. The left recurrent laryngeal nerve goes from the larynx and the trachea to the brain quite windingly for it makes a detour around the heart before ascending to its destination. This adds up to an impressive fifteen feet in the case of giraffes. Why would an omniscient and omnipotent great watchmaker bother to resort to such sophistication? In the evolutionary frame, this would naturally come about. A direct trajectory from the gills to the brain in former tetrapods could be bent by the slow shifting of the heart, provided this doesn't induce any lethal aftermath, whereas the restoration of a straight course demands unlikely simultaneous restructuring.

As a last comment about evolution, we will reformulate it with more ambiguous words. We will then try to avoid these common words that in favour of more precise terms. By examining the words good and evil we conclude that good represents life's stability at a certain scale, while the word evil summarizes the strikes against it. Selection proceeds according to the statement: "Good draws its strength from the attempts of evil." However, this brings confusion as duty and will erroneously seem to interfere with selection in the last sentence. Therefore, this shouldn't be misread as if we were inciting to willingly commit evil action, for evolution isn't based on any categorical imperative over the will!

Chapter 2 - Parting

« Der Mensch ist Etwas, das uberwunden
warden soll. Was habt ihr gethan, ihn zu uber
winden ?»

Nietzsche, Also sprach Zarathustra

« Man is something that is to be surpassed. What have ye done
to surpass man? »

Nietzsche, Thus Spake Zarathustra

Geometric thought is that effort that is conducted in the direction of finding a *reduction* of a complex form or a complex problem in such a way that that reduction is simple to study although by the insight it gives on the phenomenon it is significant. It is all the opposite of *statistical methods* that give interpretations based on a small number of pre-existing models. So, in that sense geometric thought is a demure way to view the world for it does not try to explain everything without focus on what that may cost. However, it is a very powerful tool: when such a geometric reduction is discovered, one can suddenly study a system far too composite prior to the description of the model in great depth.

Here are two results usually taught in a physics class that are geometric. The first one is a property of solid materials called *König's theorem*. It states that the kinetic energy of a system of particles is the sum of the kinetic energy associated to the movement of the centre of mass and the kinetic energy associated to the movement of the particles relative to the centre of mass. Another example of geometric reduction that laid the foundation of classical mechanics is the fact that Newton's second law of motion, that describes the movement of a material point, also applies to solids, that is, to bodies that have a defined shape. Specifically, Newton's laws predict the trajectory of the centre of gravity of a solid body when you consider the set of *all* forces acting on the system. By doing this one can *reduce* the study of the movements of a solid body - such as a fly, a pen, and even a planet or a star - to the study of a much simpler object that is one of its points.

One can investigate the *geometry* of the space surrounding us by describing a simple model that gives the *distance* between any two of its points. This is the conventional and *symbolic* meaning of the word geometry; the etymological definition of that word being "measuring of the land, the earth". Here we suggest a generalization of this definition based upon the fact that the type of reasoning triggered by the study of traditional geometry is precisely what we will call *geometric thought*. The birth of such momentum took place in Ancient Greece and that impulse then was carried out through history. It is alongside formalism one of the two fundamental pillars of mathematical sciences. From this perspective, planet Earth is a sphere while a field is a plane.

One can also attempt to describe certain historical dynamics and constraints that mankind has faced and survived in the recent or distant past geometrically. Unlike the laws of physics and the principles of evolution that form a frame to understand the world, a geometrical paradigm does not always have universal scope. It may at times be a necessary condition but can also apply only under given *initial conditions*. These conditions should then be stated precisely, and it is implicit, when one should offer a geometrical explanation for past or ongoing human historical events, that the proper initial conditions allowed these events to happen. The reason why initial conditions are often omitted when an account is made of events plainly is that we most of the time lack precise and ascertainable information. However, and this is not without consequences, there is a point in time where we have better access to initial conditions. That is present time.

An obvious question to someone observing the economic and political history of Western civilization during the past five hundred years is to understand how small groups were able in the course of this period to gather and accumulate all the *money* available on Earth, and with it all the power this money represents *symbolically*. These *banks* have acquired such unchallenged power by mastering one simple and elegant geometrical idea. *Exponential growth* is a phenomenon that occurs when the growth rate of a quantity is proportional to the quantity's current value. In mathematical terms, the value y of the *function* that measures the phenomenon and its derivative y' satisfy at any time t:

$$\frac{y'(t)}{y(t)} = a,$$

where a is a constant amount. A simple example of an exponential evolution is the following. If a disease starts by infecting one person and if after each period of a year the number of people struck by the disease is multiplied by two, then the population of sick people grows *exponentially* with time. Indeed, after a period of one year, there are $2 = 2^1$ bearers; there are $4 = 2^2$ after two years, $8 = 2^3$ people are ill after 3 years, $16 = 2^4$ after 4 years, and generally there are

$$2^n = \underbrace{2 \times 2 \times \ldots \times 2}_{\textbf{n times}}$$

people infected by the disease after a period of n years.

A crucial property of exponential phenomena lies in the fact that the measure of the phenomenon grows *much faster* than time passes. Clinging to the example of the spreading disease, after a certain - perhaps large - number a of years, the number of ill people $2\char`^a$ will be larger than $10a$. After an even larger time span $b$, $2\char`^b > 100b$. Then after c years, $2\char`^c > 10000000c$, etc ... We can rephrase this with the following mathematical statement:

$$\frac{2^n}{n} \longrightarrow +\infty, \quad n \longrightarrow +\infty$$

In other words, *multiplicative* phenomena ultimately win over *additive* ones. It is by grasping and applying this idea some years ago that a few financial institutions now concentrate such power. If in our example you replace sick people by gold coins, and you now consider a pile of gold that sees its height multiplied every year, say by a factor 11/10, which is what an interest rate does in effect to the bank's treasure, meanwhile an *employee* earns the same amount x each year, that *adds up* to older pay checks. After a period of n years, the bank will have made $(1, 1)\char`^n \times \alpha$, where α represents its initial capital. For his part, the worker only made $n \times x$. So even if the employee's salary is big and the amount of gold the bank possesses at its birth is small, that amount of gold will ultimately exceed any multiple of the earnings the worker will eventually accumulate:

$$\frac{(1,1)^n \times \alpha}{n \times x} \longrightarrow +\infty, \quad n \longrightarrow +\infty$$

The banker chose exponential growth, while the wage earner settled for an additive, linear accumulation and basically accepted slavery.

Let us acknowledge at this point that the gold and other precious metal supply on Earth is *finite*. So there comes a point in history when practically all the gold there is becomes possessed by banks. Even prior to this moment, gold was already physically controlled by banks who would *distribute* paper notes representing a certain amount of metal owned by the bank and payable to the bearer on demand. And when the distributors finally took hold of the entire world gold reserve in the mathematical process we have just explained; American President Roosevelt went one step further and banned the use and possession of gold for individuals under Executive Order 6102. On August 15, 1971, President Richard Nixon announced that the United States would no longer convert dollars to gold at a fixed value, thus abandoning the gold standard. In effect, this transforms paper money, that used to be a debt to the bearer, into a debt contracted by the first bearer of the bill. This shifting in paradigm is referred to as *fiat money*.

A second geometrical idea that applies to understanding recent human history, but also much more generally the evolution of *remembering stable forms*, is another limit phenomenon that we will now explain. A territory can be represented by a disk of radius R. Then its perimeter is equal to $2\pi R$ and the area it defines is worth πR^2, where $\pi \approx 3,14159$ stands for the area of a unit disk, that is a disk with radius 1. The perimeter of a territory measures the length of the border that needs to be defended, when the area limited by the border is equal - up to a factor - to the number of soldiers capable of fighting to keep the land safe. The ratio of these two quantities is:

$$\frac{\pi R^2}{2\pi R} = \frac{R}{2}$$

and is equivalent to R. This means defending a strip of land is easier for a large state. Thus, large groups will be the ones that tend to endure and extend their rule, while the number of these groups will decrease with time. For example, this geometrical principle explains a series of events that occurred during the last world war. France and Belgium were allies in this conflict and each had a border in common with Germany whom they were fighting. Commanders of the French and Belgian armies decided that each country would defend its own border. Following the idea we just developed, the Belgian border was not as prepared to counter an attack since the Kingdom of Belgium is a much smaller country than France. The two armies rather should have merged for the time of the conflict all along the frontline. Not doing so naturally led the German army to attack Belgium on May 10, 1940, engaging in what is traditionally referred to as the battle of France.

So large states are the ones that are selected. This phenomenon is magnified by the fact that such states also have more resources which allows them to develop more efficient military technology.

If the marches of an empire often continue to struggle for safety and expansion, the heart of a large state is subject to gradual *pacification*, which requires more careful internal governance. To do so the power structure must dictate moral standards that suit this new order. Christianity and to an even greater degree Confucianism are two sets of moral laws that rule or have ruled over vast pacified empires. They were gradually accepted as the easy option.

Harmony is the consistency one can find at the crossroads of natural instincts and cultural imperatives. In such a pacified moral world, harmony is severely tested. *Beauty* is the struggle against that lack of harmony. Nevertheless, a disruption in harmony is a prerequisite for beauty.

The far-reaching expansion of the Roman Empire and the state of peace it reached caused the majority to leave the army. This move led to the establishment of an armed elite: the nobility. That caste was organized according to the rules one can find in the army corps. Its members followed an order of authority, with local leaders and global ones - the kings. And so that a peaceful transition of power would occur between generations, the succession problem was resolved outside the family structure by an arbitrary rule: that of birth right. However, this principle leads to a lack of selection that can rapidly degenerate. For example, it is hard to recognize the rules and traditions of a military structure when you observe the way King Louis the Great of France would run and subjugate the members of his court.

Thus, human groups necessarily try to dominate the others, which tends to an overall increase in the population of groups that survive. At the same time, *harmony* in these groups becomes more and more precarious while ideas, weapons and *beauty* reach unprecedented complexity. The expansion of human societies takes place through fusion and integration of one group by another. When such growth is complete and the fusion is total, civilization becomes very vulnerable for it is then one of a kind. Moreover, harmony is now so distant and weak that ancient natural instincts gain strength once again. Death of the civilization is postponed only by *beauty* it accumulated over centuries. The power of an empire is measured by its ability to resist annihilation. And if ever the impulse driving a civilization is vigorous enough, the land it has dominion over can be extended once more to reach new territories and therefore extend the group's lifespan. Our species is living in such a finite and one world for the first time in recorded human history. So, the fate of our times is to go beyond our horizons by expanding so we can divide once again.

Since these very simple geometrical ideas allow us to understand some aspects of history quite precisely, they also carry the potential to predict future events. To grasp exactly what that potential is, one must understand the present circumstances.

The two major obstructions humanity must now face if it is to continue thriving are its uniqueness and the *finiteness* of our world. First, the world we live in does not place distance as a limit anymore because efficient means of transportation became easily accessible in the western world during the 20th century, and since then this became the case practically all around the globe. And while remoteness becomes meaningless following the reduction of distances, that shrinking helped to promote and expand the cultural and political unity that defines our global society. It also undoubtedly disclosed the narrowness of our environment. And the finiteness of our world also has startling consequences since it becomes clear now that we will soon run out of some key resources. For example, proven resources of silver, gold and zinc will be exhausted in less than 20 years.

There is nothing close to petroleum products in terms of energy efficiency, and they obviously play a crucial role in the past, present and future development of human societies. The unbelievable squandering of that critical resource we witness daily is quite disturbing. It does not indicate a clear understanding of what is at stake or the slightest predictive capability among distributors and commanders.

More generally, our whole environment is destined to change considerably through human action. We are spectators to the forest cover shrinking and to the alteration of the composition of Earth's atmosphere. The preservation of drinking water also is a major concern if mankind is to survive.

Growing demographics and an ever-continuing arms race add up to a closening horizon and a transforming environment to make humanity's defeat certain to occur soon. Of course, small groups would probably survive such an *apocalypse*, but what means of action will they have preserved, what knowledge and culture will there be to transmit from that point? What world will be left for these humans to prosper again? And the hand of death has multiple other cards to play than annihilation by nuclear weapons: disease or a meteorite impact are two of many dangers.

So, the seeming strength of our assumptive civilization should remind us that other mighty empires died out in the past, sometimes so quickly that no one inside the group was able to predict it. But in a *unified* world, the fall of civilization would mean extinction for all. A *geometrical* perspective shows us that we are destined to disappear if we do not put an end to the *finiteness* and the *uniqueness* of our territory. Unlike the Maya and the Assyrian empires however, we are not falling into a ravine with both eyes closed.

The authors can see only one way out of this otherwise compromised future we prepared for ourselves. Men must explore and conquer lands unknown to them for now - as they always did throughout history - so we can *part* what has become a universe and endure once again as a multitude through the next age. The making of different branches of humanity that can evolve and experiment independently will stop local mistakes from being lethal globally.

The meaning of this *expansion* of the human kingdom is the

quest for more resources and new spaces. Consequently, it is also indirectly the pursuit of a longer lifetime. Obviously, moving part of humanity out of its primary habitat implies conveying both bodily forms *and* knowledge. Therefore, members of such an expedition cannot be selected only based upon their physical and mental toughness.

About space exploration by man, the first question is: what were the true achievements of the conquest of space during the twentieth century?

The simple way to answer this is: "Men were sent into outer space to the immediate surroundings of planet Earth for short periods of time and have been able to walk on the surface of the moon. Many artificial satellites are now orbiting the Earth and space probes have been sent to different locations in the solar system."

Is it reasonable to rely on that account of events? Available space around Earth - such as moons and other planets - would give a people able to conquer these environments a significant advantage in terms of expansion and of newly available resources that would guarantee its long-lasting existence. We assume that any man who truly can direct the course of human history, and who must therefore have an extensive and foresighted understanding of our world, is aware of this. For the preservation of the aggregation of men he might represent, such an individual may decide, if there is a possibility for the group to extend their reach, to keep what can be hidden quiet on that subject, to stay on the path leading to victory. The question that immediately comes to mind is: "Is it possible to falsify history to that extent?" We state this criterion: a historic event that occurs in the presence of a great number of witnesses with different wills cannot be concealed. The Napoleonic wars took place. By contrast, caution should be exercised when considering such an episode where the events or their meaning are relayed by a small group of individuals with possibly converging interests. For instance, the implementation of the Manhattan Project involved many Americans all through the second world war. Still, the goal stayed secret throughout the whole development, which was rendered possible by the limited number of people who actually knew what that goal was. Similarly, rockets have been fired from locations that are well documented and observable by many, but who can say they know precisely what these space engines were transporting? The spectrum of possibilities for mankind's achievements in outer space is broader then we may imagine.

This observation is not meant to be a vulgar complaint. Lies and secrecy are at the core of human nature. This is a psychological tautology.

The goals behind this space race are not apparent either.

After he returned from his first voyage to the Americas,

Christopher Columbus convinced the leaders of Spain that they had much to gain in settling there and in colonizing these newly discovered lands by mentioning abundant gold reserves. The story of his travel was a combination of truth and flamboyance, and it made a great impression on the royal court of Madrid. For example, he explained:

> " Hispaniola is a miracle. Mountains and hills, plains and pastures, are both fertile and beautiful. [...] The harbours are unbelievably good and there are many wide rivers of which the majority contain gold. [...] There are many spices, and great mines of gold and other metals. "

Did John Kennedy have the same narrow vision that the government of Spain once had when he promised on September 12, 1962 that man would set foot on the moon before the end of the decade? Did *they* simply want to compare the strengths and weaknesses of the liberal society versus the communist system? Or else?

Yet, since we do not have a direct access to *reality*, we will base our reasoning here on the narrative that is usually taught as true.

The "conquest of space" started during the cold war with the Russians and the Americans. We can identify two significant periods in that "race".

During the first phase, the Russians had the upper hand over their opponents: the Soviet Union launched the first artificial Earth satellite, and then successfully sent the first animal and the first man to orbit our planet. Then the Americans outsmarted the Russians as they became the first nation to send a manned spacecraft to the moon. They repeated that accomplishment on several occasions before abandoning any project as ambitious as a moon landing. Nowadays, our presence in outer space is limited to an inhabited space station that continuously orbits the Earth. And apart from numerous artificial satellites orbiting near our planet, a few spacecraft are sent farther away for scientific purposes.

What can we learn from that? Firstly, political systems have a major influence on the outcome of such ambitious undertakings. An authoritarian governance enables a group to focus on a fixed goal, but the completion time must be short, for what has been neglected because of the main task results in economic depletion. A "democratic" flattering regime certainly is more sustainable, but with time it leads to a weakening of the will driving a project.

How can we evaluate the ability men believe they currently have to settle in distant worlds? An interesting sign is the recent focus of *media* attention on the risks there are that human activity destabilizes our natural habitat. Well, there is no reason for consequences such as the exhaustion of terrestrial resources not to have been long foreseen. How then should we explain that a response was so late and slow to come up with? The only logical explanation there can be is the belief now past that there was a loophole to escape. But that route immediately comes to mind: before mankind was encouraged to deliver a *mea culpa* for its actions on Earth, the goal that was set for us was to expand our horizons, to the moon and perhaps even further. But that task was suddenly forgotten. It may be that it has been ultimately labelled impossible by those in the upper reaches of power? How then should we account for interplanetary probes that continue to be sent to Mars and other destinations? Scientific interest can certainly explain this in part, but can curiosity alone justify the necessary fundraising for such a perilous and complex enterprise? Moreover, why give the priority to planet Mars, a celestial body within reach of humans, on which we have found traces of water - in the form of ice? This seems to indicate remaining hope that men can settle there. Besides, momentum could originate from the prospect of extracting abundant extraterrestrial resources for an earthly use. For example, heavy hydrogen (deuterium) can be found in profusion on the moon, and could open the road to a new American *El Dorado*.

Anyhow, it is not up to man to decide what is possible and what is not. Not all challenging and hazardous efforts end in defeat. So many experiences of the past prove that, while men have surpassed themselves over and over again. Severe stigmatization is now imposed upon the whole of mankind: violence we have demonstrated towards other species should bring *shame* upon us all. But life was never ashamed of the brutality it displayed. *A fortiori,* how not see the mistake that is being made with such disregard when it is man's *destiny* to honor and pay tribute to life by increasing its kingdom and setting up a new framework for it to flourish?

By *destiny,* we do not mean something that is necessary. On the contrary, what we call *destiny* is that *possibility,* that is encountered only *once* in the history of Life, to modify its course permanently. Destiny is not predicted by the theory of evolution and is not always attained. So few life forms have a destiny to fulfill, can humanity give up without a fight? All species disappear and mankind will definitely follow this road someday. But before that tipping point, we can participate in making what we belong to prospectively immortal: Life.

Nonetheless, we need to recognize that after our victorious landings on the moon, the attempts to further reach and settle in outer space carry the mark of repeated failure. This necessary and symbolic first stage may have captured our attention too exclusively, diverting creativity from the other issues that need to be cleared up. Perhaps, after the jubilation, commanders wondered: "What do we do now?", without being able to set the path.

Indeed, the obstacles to be overcome before we can possibly establish a human outer space colony are far more diverse than simply the transport method. What social and political forms are best adapted for a community facing hostile and largely unknown surroundings? We can certainly answer parts of that problem, but nothing close to what all needs to be planned prior to such an expedition. One important feature is the following and a consequence of life's evolution theory that sets precise stability conditions: a colony must be divided into several groups living in different locations, so that the accidental loss of one tribe does not jeopardize the whole mission. How to survive? Could farming be practical on the long term using crops brought from Earth and planted in an area specially arranged for them? Can these same plants be used to form an atmosphere resembling that which we have on our planet, in particular regarding its oxygen content? If the expedition was to go beyond our moon to planet Mars - or even further on - would the crew be capable of withstanding the physical and psychological hardships that ensue from isolation? Regarding psychological problems due to loneliness and prolonged lack of space, although it is for sure a major issue for individuals brought up in an indulgent political system, the fact remains that history displays multiple examples of much worse psychological torture inflicted on men by their peers. Consider, for example, the monastic lifestyle that was forced upon some societies by religious groups. This is nothing but self-discipline.

On Those Who Make Mistakes in the Oratory

When anyone has made a mistake while reciting a Psalm, a responsory, an antiphon or a lesson, if he does not humble himself there before all by making a satisfaction, let him undergo a greater punishment because he would not correct by humility what he did wrong through carelessness. But boys for such faults shall be whipped.

Rule of Saint Benedict, Chapter 45

Health and physical well-being on board the spacecraft are undoubtedly more worrisome questions.

Here were a few of the riddles the Sphinx of *reality* will ask of interstellar travellers, among many others we do not yet suspect. But we do not seek to give answers in this text, for we cannot. Still, we can set out what we believe is a preliminary to solve: what social and political organization would be able to focus a significant part of humankind on accomplishing that end? This system would have to forge *beauty* for that purpose. It will require a political genius to create and enforce that form that will push Man forward. Chapter 5 raises a few related points. And this is not close to being a sufficient condition. It is essential that all fields of knowledge, including hard sciences, be tested thoroughly in order to assess their fruitfulness.

The declining resources we already put emphasis on compel mankind to act fast if we wish to live up to our *destiny*. Moving part of humanity out of its natural place and developing technologies that will allow that to happen will require a gigantic amount of energy. We will not have that kind of power available forever. The day we run low will be too late to start searching for an escape route that became impracticable. We must act while a possibility still exists.

There is another limit to when this project still can be undertaken. The disappearance of colonizing civilizations we now witness through enslavement and distortion of western civilizations by their princes has already undermined the momentum generated by the *Faustian outlook* [5]. The dying of that cosmic beat, that the extraordinary surpassing operated by western man in only a few centuries originated from, is hastened by the unification of the world through forces that do not share that tension of the soul towards the unreachable.

The current centre of power and the one to come, the *world-city*, home to the symbolic power of bankers, and China, the Kingdom of conservation, have projects and founding texts that carry no trace of *consciousness* - defined as in what follows - capable of reaching forward. Neither do their actions suggest an awakening.

And if the concord of all human works puts men together to escape a world that suffers deeply from our presence, but is not enough to free us, it will at least replace older projects, shaped in a period where conflicts between cultures were a necessary consequence of evolution, by cooperation leading to greater stability.

Princes, human art has been given to you for the purpose of spawning novelty as much as for you to prove that you can turn entire lives to where you have decided. Put it to good use. Do not neglect evolution, and remember that almost all species that ever lived died off. And if all humans perish, those that you command will be no more, you will no longer govern, and all will have been in vain.

Chapter 3 - Diffusion

Currently, planet Mars lies within range of mankind. It is being carefully probed by engines sent to collect scientific data, as we are told, with much precaution not to pollute this land with earthly bacteria. What a pity, indeed, if one were to discover life brought by man! That devotion of Life to Truth makes sense in Platonism, but if Darwin is correct, what will befall human knowledge after man's extinction? Complete oblivion. Why then should man be sorry to realize life he brought has thrived. Is a parent sad to behold his child?

Human colonization of other planets would widen the scope of *consciousness* or at least of *will*. These terms we will define in the next chapters. This could entail the beginning of the *multiplication* of human shelters in outer space. However, if mankind was to concede it cannot adapt outside of earth, a new riverbed for life could nonetheless easily be dug by selecting *simple* lifeforms from earth for their adaptability and vigour and carrying them to remote planets. That would give a slight chance for life to initiate a new cycle that could itself - after a time our minds cannot predict - through a new will sow further life.

Bacteria are the most adaptable life forms on earth and mankind should logically choose among them for this purpose, by sampling them in a place resembling the extra-terrestrial destination where they will be delivered. This is possible, only will apparently lacks.

And what if terrestrial life supplanted a native life form? That is the principle of life.

One might also be concerned about the fact a bacterium that has freely evolved on another planet could accidentally return to earth and cause a disease. However, its adaptation to its new planet should not increase its virulence on earth, rather the contrary, and man should not in the first place send an infectious bacterium. Diseases are indeed dangerous for mankind, but the threat comes from earth.

There exist hostile environments on earth in which men have not yet settled: the poles, the oceans and underground. They are not used except to extract ore, and could serve as laboratories for men to train in extreme and inharmonious situations similar to what they could expect in outer space. These places might also provide the bacteria that could be scattered in space.

If it isn't fulfilled through mankind, how could such an expansion take place? Life took a long time to pierce the *surface* of the ocean but it is far easier to cross this fence than the *volume* separating earth from another planet without any resources, which no living form *close* to what is found on earth can achieve. But evolution without will only proceeds to add a *small* variation to another. Will is the only shortcut we are aware of to make this jump.

Even if this does not really matter, let us make the point that if man can transport life from earth to another planet, it is possible that earthly life has its origins outside earth.

Chapter 4 - Creation

« Darth Plagueis was a Dark Lord of the Sith, so powerful and so wise he could use the Force to influence the midi-chlorians to create life ... »

Star Wars episode III, Revenge of the Sith.

« Wird man vielleicht uns einstmals nachsagen, dass auch wir, nach Westen steuernd, ein Indien zu erreichen hofften, – dass aber unser Loos war, an der Unendlichkeit zu scheitern ? Oder, meine Bruder ? Oder ?– »

Friedrich Nietzsche, Morgenrote, 1881.

« Is it possible that people may one day say of us that we also steered westward, hoping to reach India—but that it was our fate to be wrecked on the infinite? Or, my brethren? or—? »

Friedrich Nietzsche, The Dawn of Day, 1881.

Darwin's theory of Life's evolution accounts for its durability but provides no scenario regarding its origin. *Will* cannot a priori be excluded as its initial impetus. *Will* can be decomposed into four steps occurring in that order: conception of a goal, planning of the means, and then execution of the plan and the goal.

Man has invented many efficient mechanisms, especially some that build other mechanisms. Is this not the first step towards the *creation* of life?

Figure 1: The four steps of *will*.

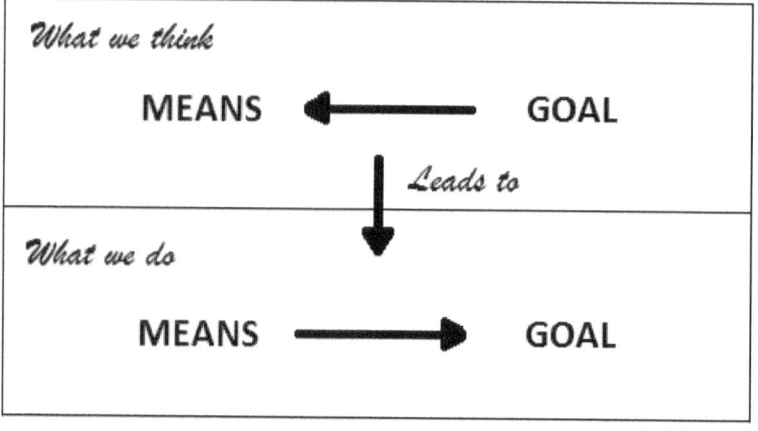

Chapter 5 - The Subject and the Object. On corollaries, not set as goals of Will.

This Chapter is devoted to analysing some intrinsic attributes of human beings that should not be overlooked in instituting a social form capable of accomplishing the ambition to conquer a land never treaded before.

The word *truth* embodies a model of *reality*, the latter being trusted as self-existent. As such, *the* Truth as an absolute doesn't make sense, and countless truths strive, championed by groups and individuals, to earn the privilege of being selected throughout evolution. Some truths are more efficient than others which is the only value that we can bestow upon them. There once was a time when Galilean truth would legitimately scare the individual as dangerous shadow and darkness next to the Church's gleaming truths.

Truths are voiced through the use of forms or *concepts*, and we would like to distinguish two types of concepts: subjects and objects. In this chapter, we intend to inquire which minds are more susceptible to each one and how these different concepts sow different *moral* seeds which do not prosper in the same context.

What is the avenue leading to an idea's triumph over a cheering crowd? In order to persist, it can't harm the idea to benefit the expansion of a group. However its explicit expression in the multitude follows its dormancy as an *intuition* in many. For none can persuade who isn't already convinced. It is then released by a respected authority which can address a large audience, due to its charisma, its erudition, its power may it be political, military, financial or even magic when prophets dispense multiplied bread. Once a substantial majority accepts the idea, the rest is easily rallied.

If one refers to *the* Truth, the history of an idea is an anecdote, not a token of its value. Aside from such idealism however, this remains a matter of concern. An example of research of the origin of an idea is Nietzsche's diagnosis about Socratic morality:

This foreigner told Socrates to his face that he was a monstrum — that he harbored in himself all the worst vices and appetites. And Socrates merely answered: "You know me, sir!"

Nietzsche, Twilight of the Idols.

That we say for ideas hatch from Subject and Object, and they are there- fore related to the *history* of thoughts.

To the *subject* belong the concepts that depict the world through human sensations and emotions.

The *object* highlights the action a form has on other forms to characterize it.

The subject is constituent of *femininity* while *virile* thinking is objective. Logic attests of this as this distinction emanates from the very mission of each sex. Let us initially neglect the moral *culture*, in other words the memome, contains and concentrate on human animal *nature*. Woman, as she bears her children is, unlike man who can leave during this period, supposed to raise them to independence, and she must be able to read *inside* their individual mind to succeed. Isn't women's amazement before children striking when, smiling, they exclaim "Isn't he cute?" Man isn't compelled to accompany his progeny. He must breed. Have the greatest possible offspring, which costs him far less than for a woman. To ensure this, he can direct his attention on the number of women with whom he matches rather than on their personality. If he stays attached to a woman however, it is also plain that he must be the one to bring back most food provision and to fend off aggressors when his wife is pregnant to keep her safe and rested. He therefore, always endeavouring to catch, defines forms as *linked* to other forms which is exactly to label them as objects. Tautologically sexes are distinguished by their role in reproduction, nonetheless, this asymmetry extends to survival mechanisms.

As a solely natural human being is fictitious, culture has a significant influence on him, so it is important to understand how subject and object mold culture. In traditional societies that bear witness of what times of yore, that were closer to nature, looked like, woman dominates *inside* her house thereby complying with her subject nature. Man is in charge of dealing with issues coming from *outside*. He nourishes his family and protects it, sometimes even by fighting at war. And objective thought as well as man's body approve that. Yet, as society of which family is a first hint, overrides nature, the boundary between male and female blurs. Indeed, society requires stability in the relation between its members, and will for instance discourage unfaithfulness. Moreover, living in a community forces to learn its custom and many usages, from the manner people interact and communicate to the knowledge of an art of which one can bargain the fruit with others. These teachings are both feminine an masculine. Hence, social stability and education are the two pillars of marriage. Its firmness resides in a feeling unleashed by culture called *love*. As fidelity towards a woman can seem restrictive to man, woman must transform into a particularly at- tractive object in order to rule over the subject. Hence the makeup, jewels, dresses, ... However the contract between sexes is never flawless and woman can fear man's violence and man woman's violence. Man's violence, as he directs it towards objects, is physical when woman's violence targets the mind.

Of course, some may say this split between sexes on subject and object is exaggerated. Yet, it has strong theoretical foundations and is appropriate in practice in many situations, as we will see. A cause that could explain how these two types of concepts come to mingle is that moral evolution can impose ideas from one sex to the other. Indeed, up to now we only considered a minimal morality, too weak to visibly inflect human nature.

The Jewish people was assembled and united by the writing of a text as a social object, symbolically represented, living among other peoples and other gods, and its religion and god cement the people as objects of social link by giving it a state and a law. This people's identity reinforced through difficult times and exile, becoming extremely acute as regards social representation of symbols and social psychology. This precision translated into a vast understanding of monetary and political representation. Clearly they did not always obey the command they had fixed as a rule : "Thou shalt not make unto thee any graven image, or any likeness of anything that is in heaven above, or that is in the earth beneath, or that is in the water under the earth." Nonetheless, inasmuch as Hebraic cultural objects are meant to integrate specific individuals in a often scattered human group, it must affect the subject, that becomes a central object. Perhaps this equilibrium redounded to this culture's exceptional longevity. More generally, let us point out that the association between object and subject produced the idea that a group of subjects, can, as they behave as objects be brought to share a will, which is the goal of glory and the mean for political power.

Grecian thought, that was established in a warlike epoch, when each people's territory was under constant siege and scrutiny by its different neighbours and changeable, was through no mystery shaped by virility and objectiveness. Objects are indeed suitable to describe these movements and confluences of peoples inside a natural world, where nature itself plays a part in man's life through seasons on which harvests depend, storms that break boats, etc. In such circumstances, balances of power prevail. Life faces time and change. An ancient Greek man exists as an element of his city, and at another scale of Grecian blood, temperance keeps him to this rank. He can purchase slaves and some may care about them no more than tools. Another key to the spirit that guided ancient Greece lies in their divinities. They personify different aspects of nature with their symbols, their powers, and of humanity with their personality, as well as the city they are in charge of. This objective thought applies to common Greek morality, but one may be slightly more specific about philosophy. *Cynicism* magnifies objectiveness as if it were the only eye to gaze at the world. Yet Greece also gave birth to far more ambivalent philosophies. Socratic morality, as Plato transcribed it, detaches from Grecian orthodoxy. Indeed, it is to prevent escalation of impulses inside the individual that Socrates recommends to search for the Truth. As it is infinite, it can satiate any desire thereto linked. And although Socrates taught only to seek the Truth, he aspired to control the minds of those he met in the city. This reveals both objectivity and subjectivity. Whatever Plato may say, Socrates was corrupting Grecian tradition and in this perspective, he was condemned in justice. Socrates made a strong impression on Plato, who also was reluctant to share Grecian thought but developed a system distinct from Socratic morality. He uses concepts that fit neither the definition of objects, nor of subjects, which he named *Ideas*. For if they are not feeling subjects, they exist by themselves independently of anything else, so they are not objects. They can, we believe, barely be grasped by man, for they do not commensurate with him. However, it may be a theoretical frame in which to place some of man's intellectual activities, such as mathematics. If one doesn't admit *the Truth* of the (logical) axioms founding mathematics, it is startling to realize how merely formal reasoning efficiency (though guided by geometry) doesn't dry out far away from these axioms, and *nonetheless* keeps producing consistent results. I am not saying that evolution cannot originate logic, but its performance even where it was not challenged by selection is remarkable. However this means it is a powerful instrument, and we won't speculate any further on a hypothetical metaphysical explanation. Plato also wanted to hold sway over politics. He even suggested that the true achievement of a city would only come from its giving rise to a philosopher king. Aristotle, although Plato's student, radically came back to Greek judgment. In his writings, the city has no duty to serve individuals. His perception of friendship also proves his objectiveness, as he saw it as a tacit contract between men whose length varies according to its strength.

Western civilization is made of a far different metal. It comes from the Roman Empire, that always tended to expand through war but also assimilation which caused it to transform in the process of adapting to its population. When Christianity first came to light and grew, *peace* within the Empire turned out to be a crucial factor. Indeed peace allows man to retreat from war and his action on the world. Easiness entices him to make that choice. To woman, that may constitute a loss of virility in man, compensated however by the stability her family gains with a more calm and long living father, whose masculinity she can detect in verbal contests of conviction rather than in physical confrontation. For political leaders this is an opportunity to avoid social troubles. Christianity, this fitting spark of genius, altered man's weakness into greatness, favoured by God, and promised to avenge oppression after death. Of course, this cultural mutation happened in-homogeneously over centuries but man gets used to peace and if women wish to bring him down he cannot compete to reproduce. He must, as a victim of feminine violence on his subject, admit as a truth what he knows is a lie, that modesty and meekness are illustrious. And the justification for this is obviously expressed by femininity: objective violence is turned against a subject, another self. The trap is well hidden in a disguise of tales invented by feminine civilization praising the harmony of sharing love, as a reward for abiding by moral conduct. That delusion conceals love is a struggle between two forces directed towards the other's subject. Still, among soldiers, victory's honour lasted, which begot nobility but what Nietzsche afterward called "slave morality" became manifest. The final stage of Christian morality is Kantian morality. The latter only abandoned its myth and admits the rest. Morality becomes necessary. *Another* cannot be the *object* of intended violence, for universal man does not separate himself from other subjects : "Act as if the maxims of your action were to become through your will a universal law of nature." The objectiveness of Kant is denatured to such an extent that he speaks of the maxim of the action rather than of the action itself. To practice *universal subjective* morality, one may submit to such a precept as: "I am a subject, the subject is universal, therefore morality consists in applying my will concerning myself to everyone else." A more evolved doctrine, professed by some who would boast of a greater wisdom reads: "from my will concerning myself I can extract principles such as freedom, equality,... These principles must govern my will as it faces universal subject." Ironically, these principles of liberty, equality, are nothing else than jealousy of stronger forms, capable of more action. They are purely selfish feelings but as they are decorated in these words and generalized, their content of pride becomes almost invisible. However, for man to be incorporated in society, a seeming allegiance to such dogmas is compulsory. As the actual belief in universal subject pertains to the inner being, it can be dispensed with, although it most certainly consolidates the will in order to conform with morality's consequences. When a morality of symmetrisation in the purposes of the will concerning oneself and others goes without that sense of belonging to a universal subject, we will call it *pragmatic*. When it is a morality of principles that disregards universal subject, we will qualify it as *hypocritical*. French thought for instance amply relies on hypocrisy. In both cases, a return to objectiveness is thus operated. In practice how- ever, these moralities condemn objective violence. At this point, it seems useful to list what western civilization deems the worst crimes. First of all, murder which is the most severely repressed, and is divided into two categories: objective murder, which has as intermediate goal to eliminate the influence of a certain individual in the world, and subjective murder which is meant to avenge oneself by annihilating the subject. Let us notice that murder has never been entirely forbidden by law. Warriors even murder as a duty and law itself can resort to murder in the form of a verdict favouring the death penalty. Society also happened to tolerate murders that were not decided by such authorities, like in the case of duels. The latter however, have now completely been banned in the West. That does however imply that no distinction is made in the law between objective and subjective murder, but how could the inside motive be ascertained from outside? Other crimes the West sees as infamous are mostly objective and include rape as the archetype of man's violence against woman; and theft, which is unlike the two last examples defined as a rejection of society through its establishment of property, nonetheless its relative importance among crimes is itself significant.

More surprisingly, new laws are emerging to prohibit violence against the subject such as racism, homophobia, harassment, ... They are no indication of past morality, but rather marks of a *will* from a power resolved to tamper with slave morality. In which direction? We cannot say. Indeed, every distortion in society provokes a spontaneous repulsion or approval in those who are not prepared to analyse its outcomes carefully. Which of the two reactions is expected, if it is not both, to bring disturbance, is hard to guess. However, it is plain that the possibility of such a reorientation of morality results from universal subjective morality based on principles. But offering such change could only be done after the topology of thoughts and men of old was broken. For Western central power now has the means (the media) to preach as one voice to the whole people, and maintain its population in a state of everlasting geographic motion and mixing that permits quick circulation of empty memes but hinders the transmission of any complex theory.

But we are diverging from the question, let us come back to it. Alongside subjectivity, one of the most successful components of Western thought, its science, is highly objective. Leaving

71

aside mathematics, which study objects initially conceived as close to reality, but that subsequently interact not in reality but in a fixed framework of logic, Western science discusses experiment, in other words the observation a man does of what *links* him to the outside world. He will then try to simplify his perception and interpret exterior forms as connected to each other in a *model*, thereby demoting the observer and subject from its central position. The models then must resist reality assailing it through experiments, which provide criteria of selection and evolution in science. Western science has bloomed since Renaissance, and has always contended with church, on now obvious grounds, as testified by Galileo's trial. Science has now reached an understanding of objects unbelievable only four centuries ago. But what could have become virility's revenge turned out to be its humiliation. Science supplied its discoveries in what became known as *progress*. This made man's bodily strength almost superfluous in the effort to supply necessaries and economics nowadays resembles a broad stage where men spend time and tire as they are confused and disorientated by women, who are their queens. A clear symptom of science's objectiveness is that most people who follow its method are men. Only sciences that appear to focus on subjects such as medical sciences or biology, generally appeal to women. Although the same objectiveness as in physics, chemistry, mathematics could apply to theses sciences, they do become less so as women are repugnant to this cold thinking. That adds to the fact that even with pure objectiveness, biology would have to consider far more complex objects than the sciences we first mentioned. Still, biology's lacunar results are motivated by progress's justification through medical science. In these conditions, the understanding of the links between objects is neglected and replaced by the point of view of a unique subject, namely the experimenter. This is typically emphasized by the use of tools such as statistics, that produce empirical laws, without any preoccupation about underlying processes, as the subject does to integrate in the world. On the contrary, if a man were to choose the steep path of learning sciences' teachings in the necessary order, not that he should wait to start any of them, but he cannot master one without it resting on its prerequisites: mathematics and *then* physics, followed by chemistry and then biology at all scales, and psychological sciences that treat a subject as an object, among which neurology and rhetoric, and finally history in its diverse aspects: struggles between states, politics, economics, laws, moralities, philosophy, well we say this man would acquire more potential to act than any great man of the past. And if he managed to transmit, his work would impact many centuries. But is such knowledge accessible? The obstacles to its fulfilling stand as Titans, and the most fearful is called Time. Such a man would begin his journey full of respect for the morality of his forebears, thereby gaining discipline and entering society that he will need during his formation. But at some point, he must feel unbearably weak in his situation and will decide to put himself first and to leave aside his moral devotion and his service to society. He will then need time to mature, always persevering in his will.

One may suppose that a spirit in which the type "free spirit" is to become fully mature and sweet, has had its decisive event in a great emancipation, and that it was all the more fettered

previously and apparently bound forever to its corner and
pillar.

Nietzsche, Human, All Too Human.

Even if such a singular individual could germinate and
blossom, will he still have, at his age, enough time and will to
pass on the understanding he has obtained over years to
successors? And since he did not arise from tradition, nothing
guaranties he will find how to transmit what much luck gave
him. Therefore Seneca is mistaking when he begins one of his
books with this criticism of Aristotle: "It was this that made the
greatest of physicians exclaim that "life is short, art is long"; it
was this that led Aristotle, while expostulating with Nature, to
enter an indictment most unbecoming to a wise man — that, in
point of age, she has shown such favor to animals that they
drag out five or ten lifetimes, but that a much shorter limit is
fixed for man, though he is born for so many and such great
achievements." The problem of time is all the more worrisome,
since man is now running out of it, as we will recall to
conclude.

To come back to our point, we have drawn the picture of a
West in which feminine morality commands, not suffering
humans to be perceived as objects, but that nonetheless
employs objective science to *progress*. If man was perverted by
morality, woman also was affected by objectiveness through
the scientific education she now shares with man, and by
working in the same way as only men used to. The most
vulgar expression of this phenomenon is feminism, when it is

sincerely affirmed by women, as it is displaying pride in being defeated, after having long reigned. This equalization and mixing signals a loosening in stable evolutionary forms and a slow decay leading to death for civilization. How long can feminine morality survive in a society where women behave like men? More generally, what those modifications will entail in the future relation between men and women is hard to predict.

The last civilization in chronological order we shall examine is Muslim civilization. Islam is based on pillars that undoubtedly put objectiveness forth, as they regulate the exterior action of the believer. Let us not forget what Islam promises paradise is made of, with its houris and exquisite food. Certainly a Muslim would see the outside world in terms of objects. However, the Quran is written for the individual to whom it promises either the paradise or threatens him with the worst possible vengeance from god. In fact, the individual is located at the center of Islam, and what is outside the center are objects. The individual is left quite alone.

What about the future? Paradoxically, we claim the more we venture into the future, the easier it is to foretell. For the near future is hidden behind the alteration in the space in which men live and move, and in which ideas circulate at a velocity not even comparable to what it used to be at a time not long since gone. Therefore, with no observations available, no scientific theory can determine morality's next evolution. But if we skip what is about to unfold, at a time far nearer than a century, Earth will not be able to dispense resources to men the way it now does. And what is for now a technically skilful civilization will be compelled to remember the necessity to fight for the fruits of this world. Objective thought will then predominate and subjective morality will vanish. Will this take place through an organized war between currently existing nations menacing to destroy mankind, or by civil dissolution, that we cannot say. But this transition is the moment when Man will have the possibility to fulfill its *destiny*. Tensions will then be at their highest and if sufficient subjective morality subsists, it will enable increasing objective powers to *concentrate* on mankind's survival. That will be when mankind learns if it has the strength to go beyond its earthly bonds.

Chapter 6 - The immortal Soul

What makes the goal we have been advocating incomparable to any other resides in its being simultaneously possible, yet with no foregone conclusion, and significant for the whole of mankind. We are not aware of any other project that would fully satisfy these three conditions. This could justify its choice by a man as both reasonable - who would therefore seek to accomplish some- thing possible - and powerful, which would enable him to propose a compass to mankind. But without the understanding of the possible in evolution, a powerful man could easily be misguided towards another temptation to amend Life that has never stopped to torment mankind. We mean the pursuit of unrealizable forms of immortality.

First, what are the different forms of immortality? They correspond to each form of *consciousness*. To define this word requires to find what is underlying beneath its numerous uses. Of course a conscious form must be able to sense a unity in itself but more specifically we will call conscious a *form willing* to preserve itself. Will cannot go deeper into its final purpose in the sense that will, as any arm adopted through evolution, is meant to preserve its bearer. In many cases however, will only indirectly protects a living being, as the goal of our will unintentionally entails our survival. Do we eat because we want to subsist, or rather to satisfy our hunger or our appetite?

Consciousness directly seeps from the fountain of which all actions are by-products. We do not know of another conscious form than one including a human brain endowed with a complex memome, although we do not affirm it does not exist (provided unidentified forms exist, which is a metaphysical question we won't try to answer). The atom of consciousness is itself quite elaborate as the symbiosis between genes and memes in human beings authorizes freedom in the range of forms that can be called self, that consciousness tries to maintain. A man can wish to see his lineage blossom, thereby esteeming the value of the body, or also pursue glory, or else educate his children or other pupils with his doctrine. This doctrine that includes his concepts and his will - in which one can distinguish two types of penchants, those that are sudden which we name pleasure, as opposed to happiness that implies constancy- structures a memome, or soul, to which most memories do not belong since they will not be inherited. Everyone does not possess a memome, for a memome means a set of memes organized in a consistent manner. The possibility of *an immortal human soul* consists in its transmission.

In order to achieve immortality of the soul, the body to which it is attached needs not live forever but only long enough to relay it to another. Through which vehicle can the soul be conveyed? A book written for future generations can insure conservation, *but* the teachings may be mostly oral. The book should only be given after the end of the teachings. The book's architecture could be ramified, with a theoretical stem and more practical branches. Far from the trunk, each generation should retouch it to adapt to history, but the core is supposed to be fixed. If pedagogy can be improved to waste less time and avoid superfluousness, a deeper soul could be reached.

Invention is done by imitating a form that functions in a situation in a similar although different context. The use of some mechanisms of genetics to apply them to memes that is attempted here is an example of such invention. However the comparison holds only partially, since *will* acts as a *selection* on memes, but not on genes.

In the task of perpetuating the soul, *order* matters. Here is one suggestion:

1. Sciences: Mathematics → Physics → Chemistry → Biology → History/Law/Economics → Psychology;

2. Survival: Health, defending oneself and attacking;

3. Reproduction (body and soul);

4. Will.

Consciousness can also represent a complex will concentrated in a single mind to stand for a human group. Plato's tale of Kallipolis's philosopher king [4] reveals such consciousness. The dream we wrote about in this book also should reveal such will.

The fear we mentioned at the beginning of this chapter is that potentially powerful men are distracted by another consciousness that shall always fail. That is the hope to become immortal as an individual. Besides the fact most human beings feel their memories are priceless to define themselves, individuals desert their lineage as they see their perspective as unique which convinces them they cannot be replaced. Galilean science contradicts this, and this discrepancy between what humans feel as subjects and know about objects in a world in which no subject stands out constitutes science's greatest flaw. What it could manage with objects contrasts with its failure to do so with the subject. The individual, puzzled by the difficulty to imagine several spectators of the world, wonders if he might not be alone, and if his end would signify utter nothingness. Hence the wish to prevent such a disaster. This eternal life fantasy can expect medical sciences to answer its prayers, but every specific form such as a body will one day or another be accidentally destroyed. It also has explored far more exotic solutions, such as that of transposing human thought inside a technology capable of calculating its mechanisms. This answer to *individual* death seems absurd to me. It contradicts itself in many ways, each of them being sufficient to reject it. A man might want to avoid death for he does not believe another can occupy his central position. But how then could a form much more distant from him -that has another relation with the world- do so? Moreover, machines are mortal bodies. And finally, to copy a form requires measurement, which the complexity of a brain does not allow. To sum up, some men want to transfer into a mortal and altered form another elusive form to make it immortal. If however, this hides the desire not to lose one's memory, which isn't part of the immortal soul, that would display ignorance, as a book would do no less than a computer.

Those who theorize such a future overestimate its likelihood, for if their reasoning is geometric and therefore elegant, which says that as computers' power increases exponentially, they will soon be able to successfully simulate human brain. However this does not take into account the depletion of earth resources that will eventually hinder technology's development.

We deplore the existence of such prophecies since they divert individuals that could highly contribute to mankind otherwise, more than for the danger they do represent for the human species if they were to come true. We must point out that that does not contradict what we wrote in chapter four, as machines do not necessarily need extremely complex pattern recognition to acquire stability and the ability to reproduce. These machines could be designed to conquer other planets but if such life was created on earth and competed with human beings, for example if it were to arise as a military object, it could certainly threaten mankind.

Currently, such transhumanism does not yet have a major influence. However, will directing societies does tend to unify them and kill vital processes of biological reproduction and cultural transmission, in a seemingly spontaneous and global generation of measures restricting population growth such as legalizing abortion, encouraging contraception or destroying traditional family structures through promoting homosexual marriage and generalizing women's work. On the other hand education by the State breaks cultural lineage, and is now mainly intended to prepare people for a job.

To resist this perversion of human nature, tradition is a rampart.

Shall this forever stay a goal not to be reached? A quest for a mere mirage? Maybe. But can't we discern more than poetry in Cyrano de Bergerac's *Voyage dans la Lune*? However, Edmond Rostand, inspired by his life centuries later in a play he wrote, seemed to see this as nothing else than a provocative tale. Time is a long trail with unexpected turns. As it now brings us close to a cliff, we must not waste it and build a bridge to a more secure future.

Competition against one's fellow living beings is constitutive of life, not to admit it would be dishonest. But the day looms, when mankind's survival will rely on no firm basis. To later win against one's enemies, there is no choice but to cooperate for the time being. That does not mean one forgets his identity. Life must only wait for the appropriate moment to deploy once again. But now, to show the greatest possible respect to life, it is necessary to shackle it. The challenge can be met for it is *one* and will depend on a small number of men, *individuals*. The heirloom of evolution influences but does not oblige individuals.

Appendix

Einige Thesen. – Dem Individuum, sofern es sein Glück will, soll man keine Vorschriften über den Weg zum Glück geben: denn das individuelle Glück quillt aus eigenen, Jedermann unbekannten Gesetzen, es kann mit Vorschriften von Aussen her nur verhindert, gehemmt werden. – Die Vorschriften, welche man "moralisch" nennt, sind in Wahrheit gegen die Individuen gerichtet und wollen durchaus nicht deren Glück. Ebenso wenig beziehen sich diese Vorschriften auf das "Glück und die Wohlfahrt der Menschheit," – mit welchen Worten strenge Begriffe zu verbinden überhaupt nicht möglich ist, geschweige dass man sie als Leitsterne auf dem dunklen Ozean moralischer Bestrebungen gebrauchen könnte. – Es ist nicht wahr, dass die Moralität, wie das Vorurtheil will, der Entwickelung der Vernunft günstiger sei als die Unmoralität. – Es ist nicht wahr, dass das unbewusste Ziel in der Entwickelung jedes bewussten Wesens (Thier, Mensch, Menschheit u. s. w.) sein "höchstes Glück" sei: vielmehr giebt es auf allen Stufen der Entwickelung ein besonderes und unvergleichbares, weder höheres noch niederes, sondern eben eigenthümliches Glück zu erlangen. Entwickelung will nicht Glück, sondern Entwickelung und weiter Nichts. – Nur wenn die Menschheit ein allgemein anerkanntes Ziel hätte, könnte man vorschlagen "so und so soll gehandelt werden": einstweilen giebt es kein solches Ziel. Also soll man die Forderungen der Moral nicht in Beziehung zur Menschheit setzen, es ist diess Unvernunft und Spielerei. – Der Menschheit ein Ziel anempfehlen ist etwas ganz Anderes: dann ist das Ziel als Etwas gedacht, das in unserem Belieben ist ; gesetzt, es beliebte der Menschheit so wie vorgeschlagen wird, so könnte sie sich daraufhin auch ein Moralgesetz geben, ebenfalls aus ihrem Belieben heraus. Aber bisher sollte das Moralgesetz über dem Belieben stehen: man wollte diess Gesetz sich nicht eigentlich geben, sondern es irgendwoher nehmen oder irgendwo es auffinden oder irgendwoher es sich befehlen lassen.

Friedrich Nietzsche, Morgenröte, 1881.

Some Theses. —We should not give the individual, in so far as he desires his own happiness, any precepts or recommendations as to the road leading to happiness; for individual happiness arises from particular laws that are unknown to anybody, and such a man will only be hindered or obstructed by recommendations which come to him from outside sources. Those precepts which are called moral are in reality directed against individuals, and do not by any means make for the happiness of such individuals. The relationship of these precepts to the "happiness and well-being of mankind" is equally slight, for it is quite impossible to assign a definite conception to these words, and still less can they be employed as guiding stars on the dark sea of moral aspirations. It is a prejudice to think that morality is more favourable to the development of the reason than immorality. It is erroneous to suppose that the unconscious aim in the development of every conscious being (namely, animal, man, humanity, etc.) is its "greatest happiness": on the contrary, there is a particular and incomparable happiness to be attained at every stage of our development, one that is neither high nor low, but quite an individual happiness. Evolution does not make happiness its goal; it aims merely at evolution, and nothing else. It is only if humanity had a universally recognized goal that we could propose to do this or that: for the time being there is no such goal. It follows that the pretensions of morality should not be brought into any relationship with mankind: this would be merely childish and irrational. It is quite another thing to recommend a goal to mankind: this goal would then be something that would depend upon our own will and pleasure. Provided that mankind in general agreed to adopt such a goal, it could then impose a moral law upon itself, a law which would, at all events, be imposed by their own free will. Up to now, however, the moral law has had to be placed above our own free will: strictly speaking, men did not wish to impose this law upon themselves; they wished to take it from somewhere, to discover it, or to let themselves be commanded by it from somewhere.

Friedrich Nietzsche, The Dawn of Day, 1881.

Bibliography

[1] Charles Darwin. On the Origin of Species by Means of Natural Selection, 1859.

[2] Richard Dawkins. The Selfish Gene, 1976, 1989.

[3] Plato. Meno.

[4] Plato. Republic, Book V.

[5] Oswald Spengler. Der Untergang des Abendlandes, 1918, 1922.

Authors & Acknowledgements

Mathieu and Michael Miloud

« The authors would like to thank NK very warmly for her kind help and for allowing them to use her artwork to illustrate the cover of this book, as well as Hubert for his sound advice, and for pointing them to the right direction in times of doubt. »

Publisher

Four Freedom Publishing Services

UK – Ireland – Canada – USA

www.ingramcontent.com/pod-product-compliance
Lightning Source LLC
Chambersburg PA
CBHW020601220526
45463CB00006B/2394